School Publishing

Scope This Out

PATHFINDER EDITION

By Juan Quintana

CONTENTS

Scope

This Out

The Hubble Space Telescope circles the Earth, looking deep into space. The amazing pictures it sends back help us learn more about the universe.

For about 400 years, telescopes have been used to learn about space. The telescope changed people's understanding of places beyond Earth.

But the story of telescopes really started with eyeglasses. Let's scope it out.

How Glasses Work

What is light? It's a form of energy that people can see. Light travels in a straight line until it hits an object. If the object is clear the light rays will travel through it, but they will be bent or **refracted**. If the object is not clear, it will absorb some of the light. The rest of the light will be reflected, or bounced off, the object.

Your eyes use light to see. Built-in lenses in the eye bend the reflected light rays. This focuses the rays, and they form a clear image at the back of the eye.

Sometimes a person's eyes don't work correctly. Then the person can't see very well. An extra lens can help the person to see better.

Some people are **farsighted**. They have trouble seeing things near to them. Eyeglasses with convex lenses help them see better. Convex lenses bulge out.

Some people are **nearsighted**. They have trouble seeing things that are far away. They need eyeglasses with concave lenses. Concave lenses curve in.

The idea for the first telescope may have started with eyeglasses. About 400 years ago many European cities had shops selling eyeglasses. These shops sold concave and convex lenses.

The first telescopes were made from these lenses. They were tubes with a convex lens at one end and a concave lens near the eye. When people looked through the telescope, objects that were far away looked much closer.

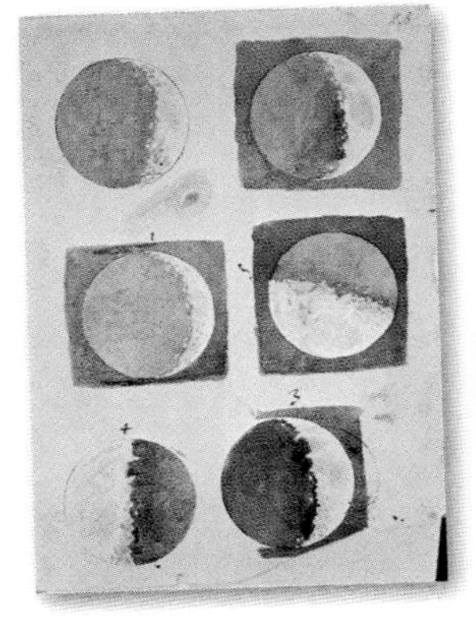

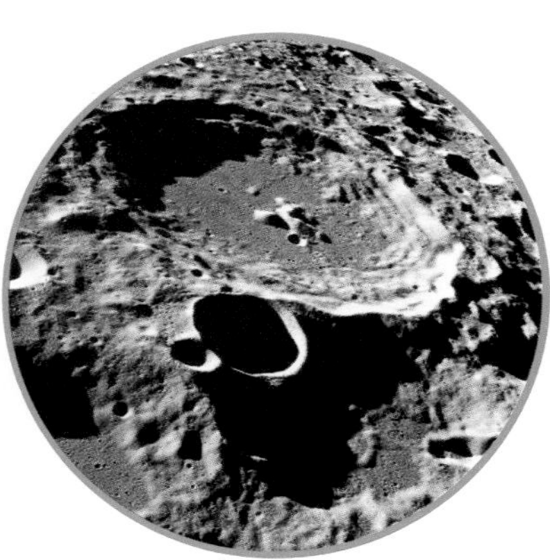

Full Moon. Galileo's drawing shows how he saw the moon. Today's telescopes show much more of the moon's surface.

Discoveries with the Telescope

Galileo Galilei was an Italian scientist. He heard about the telescope. He bought a few different types of lenses. After a while, he put a convex lens and a concave lens in a tube and made his own telescope!

Galileo was interested in what he could learn about objects in the sky using this new invention. He made his own lenses. He found the right ones to make a better telescope. Then he aimed his telescope at the sky.

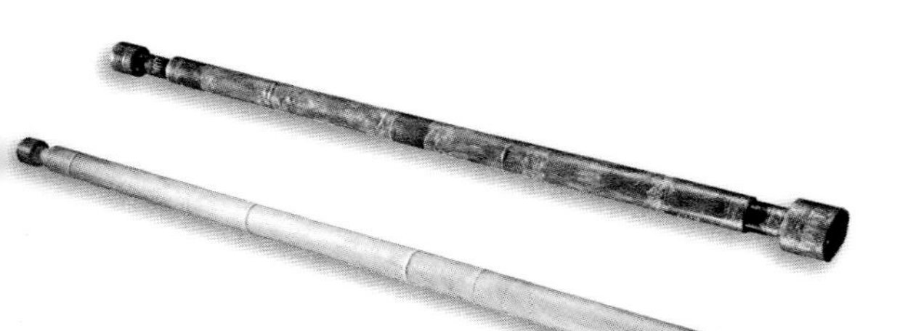

Improved Model. These telescopes were made by Galileo. Galileo's best telescopes made things look more than 30 times bigger!

What Galileo Saw

In Galileo's time, most people thought that the sun, the moon, the planets, and the stars were perfect **spheres**. They thought that these spheres all revolved around Earth.

Through his new telescope, Galileo saw that the moon was not a perfect sphere. It had bumps and holes and all sorts of scars.

Galileo also looked at the planet Jupiter with his telescope. He noticed some spots moving across the planet. These smaller spots were **satellites**, or small objects, that circled Jupiter just like the moon circles Earth. What Galileo saw helped prove that not everything revolved around Earth! His discoveries changed what people thought about the solar system.

More Power

There are different types of telescopes used to view our world and beyond. Telescopes that use only lenses are called refracting telescopes. You can make a refracting telescope more powerful by using two lenses that curve in. The first lens makes a big image. That image is made even bigger by the second lens. These two convex lenses together create an image that looks upside down!

Mirrors, Not Lenses

Some telescopes don't use lenses. They use curved mirrors instead. Telescopes that use curved mirrors are called reflecting telescopes. Curved mirrors reflect, or bounce, light rather than refract, or bend, it.

Mirror, Mirror. This reflecting telescope is at the Beijing Observatory in China.

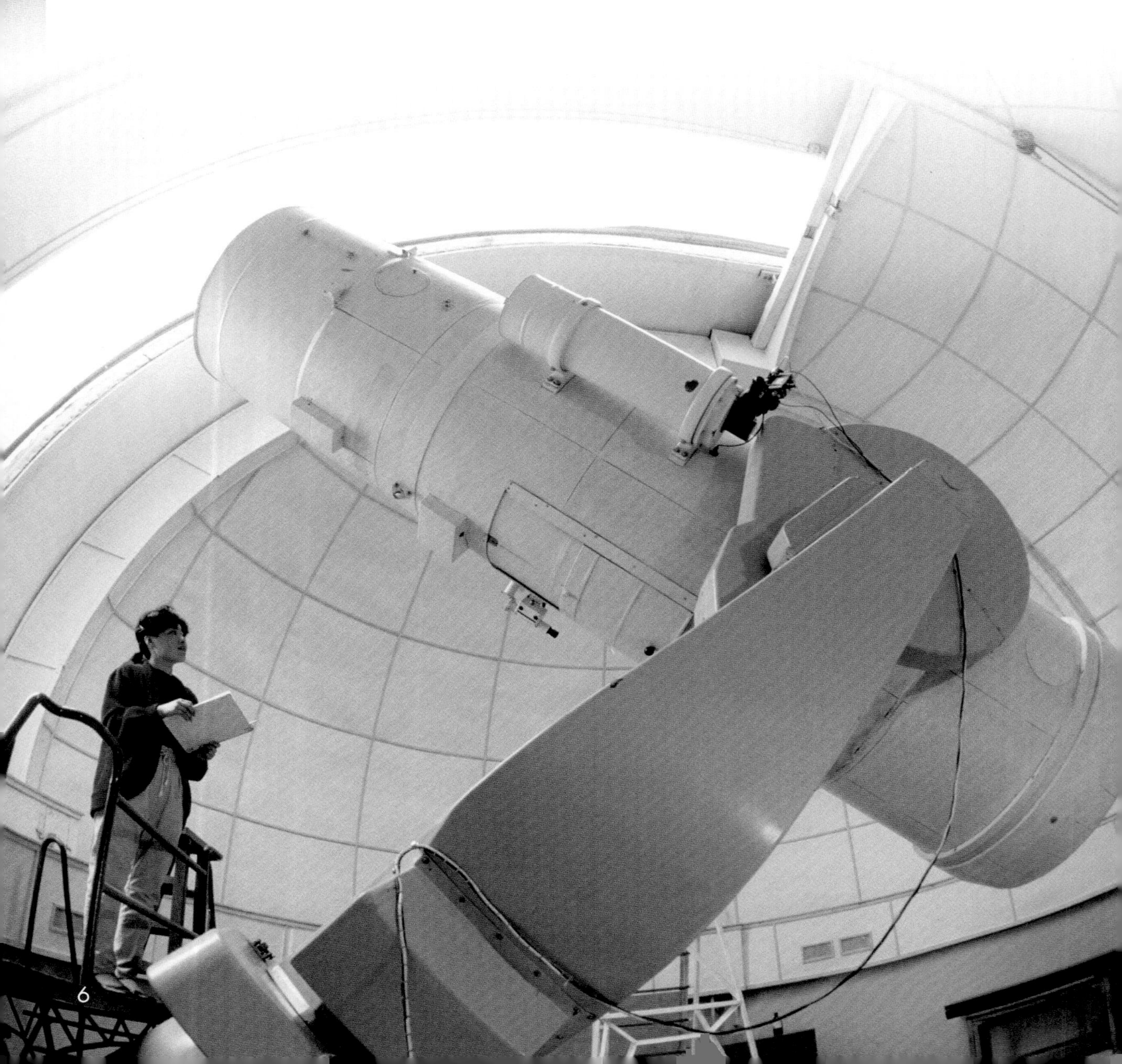

Today, the world's biggest and most powerful telescopes are reflectors. That's because with modern technology it is easier to make very wide mirrors than very wide and thick lenses.

The Hubble Space Telescope is a reflecting telescope. It floats above the atmosphere in space. You can get a much clearer image from above the atmosphere.

Telescopes have changed a lot over the years. Now telescopes such as the Hubble help us see farther into space and make new discoveries.

Looking Out. The Keck I and Keck II telescopes in Hawaii are reflecting telescopes.

Wordwise

farsighted: sees faraway objects better than objects up close

nearsighted: sees close-up objects better than faraway objects

refract: to bend light

satellite: smaller object that circles around a larger object in space

sphere: a perfectly round figure, like a perfect ball

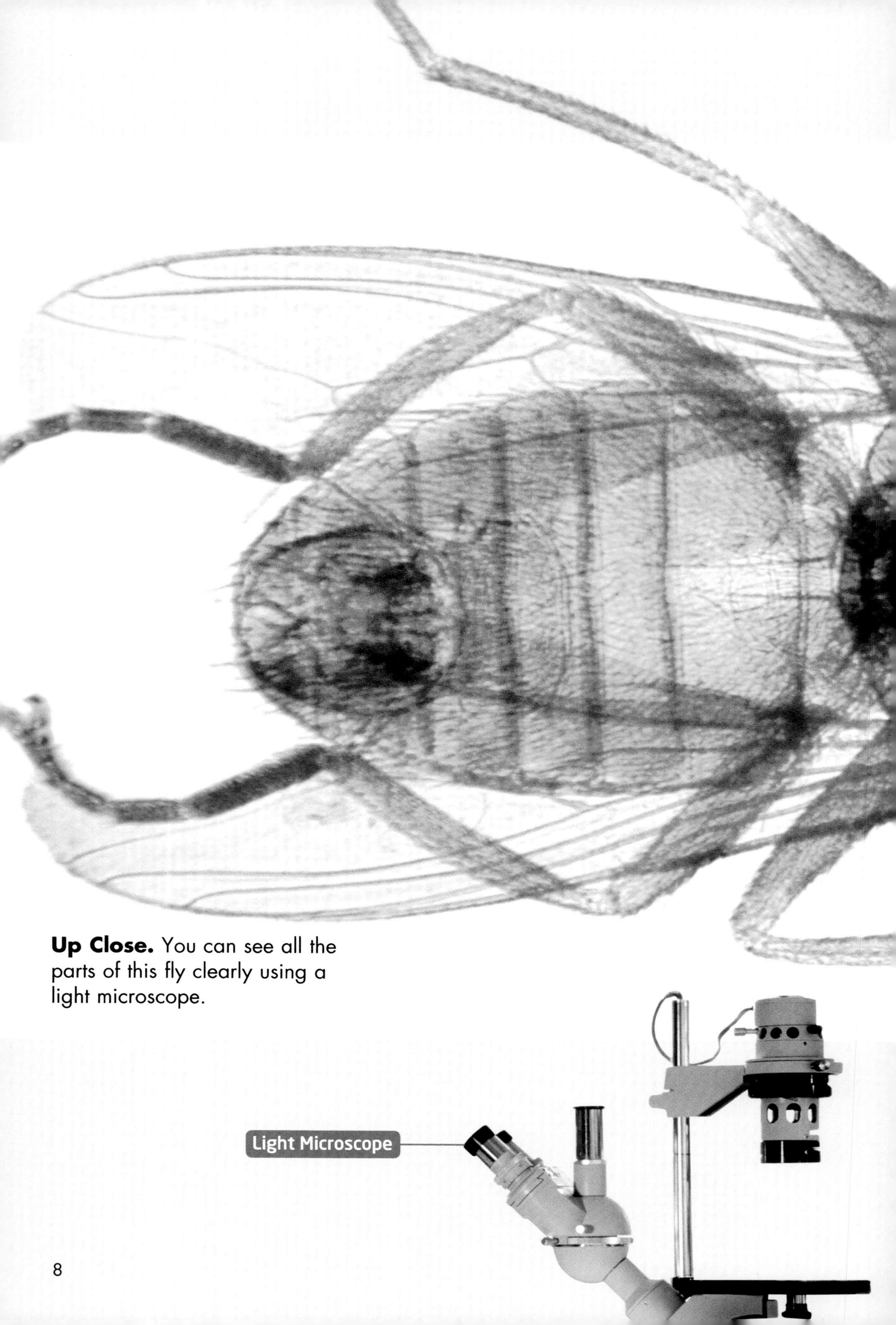

Up Close. You can see all the parts of this fly clearly using a light microscope.

Microscopes: Making Small Things Big

Imagine looking out your window and seeing a 10-foot-tall fly staring back at you.

It might be a little scary at first. But it could also be very interesting. You could see exactly how a fly is built.

Microscopes have made it possible for scientists to study flies and other insects in much more detail. Microscopes have also allowed scientists to discover things much too small to see with your eyes alone.

The First Microscopes

Some microscopes work in the same way as a refracting telescope. They have convex lenses that work together to create a very large image.

These two lenses make small images seem bigger. The bigger image made by one lens is made even bigger by the second lens. For example, one lens makes things look 100 times bigger. The second lens makes things look 7 times bigger. So looking through both lenses will make things look 700 times bigger! Here is how the math works: $100 \times 7 = 700$.

Portrait of a Flea

The first microscopes were invented 400 years ago. They were known as "flea glasses" because some people used them to look closely at fleas and other small insects.

One scientist, Robert Hooke, looked at fleas and insects. But he also looked at other things through the microscope. He drew pictures of what he saw.

Hooke's drawings and notes were published in a book called *Micrographia*. The book was an instant bestseller. It made people interested in the world of things too small to see.

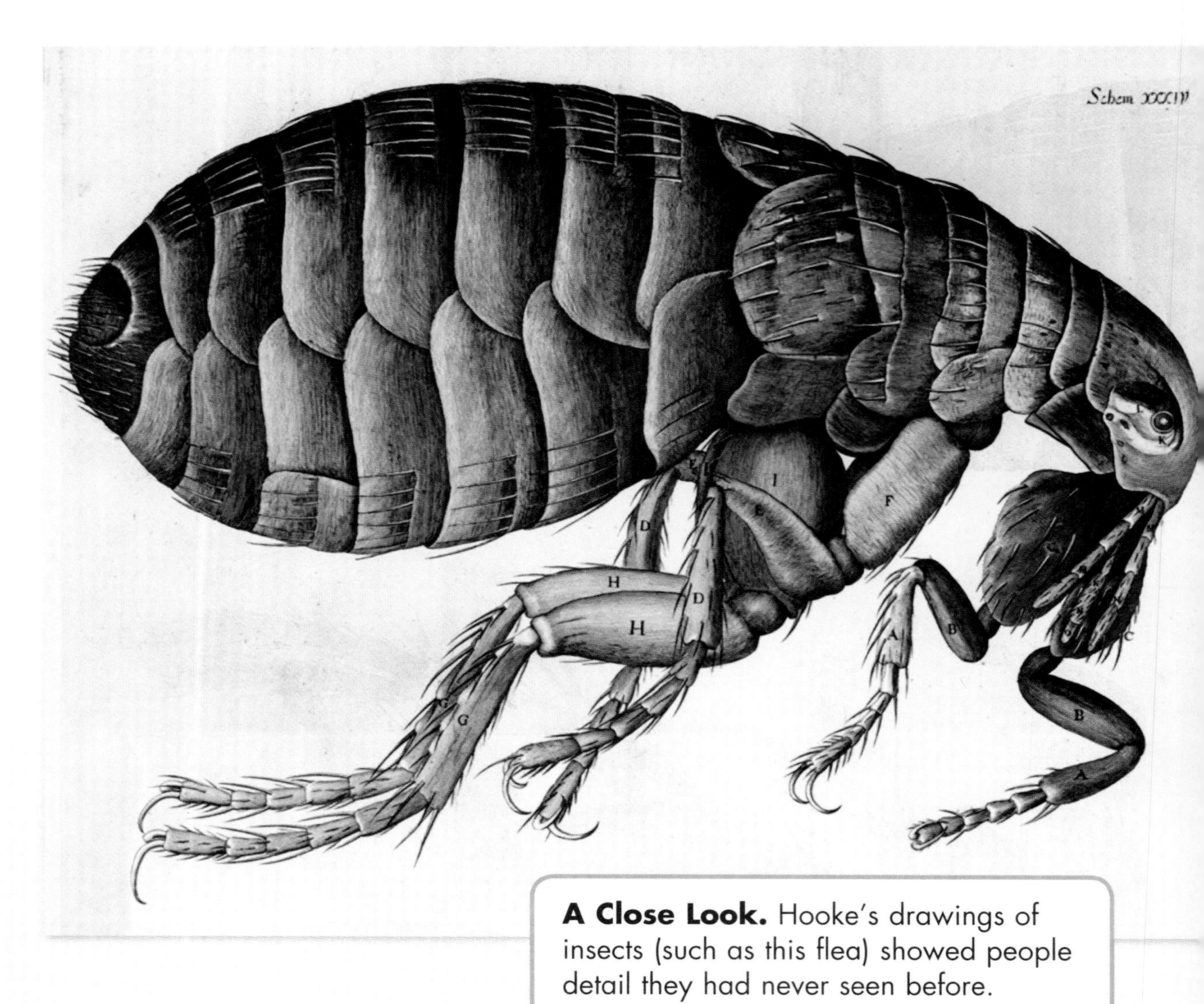

A Close Look. Hooke's drawings of insects (such as this flea) showed people detail they had never seen before.

Microscopes and Health

As microscopes improved, scientists started seeing tiny animals and plants. These were much too small to see with eyes alone. Using a microscope, scientists saw that these tiny specks were really animals and plants. And they were alive!

Scientists called these new creatures *microorganisms*. Over time, thousands of new species of life were discovered.

Scientists discovered that some microorganisms caused diseases in animals and people. They called these "bad" microorganisms *germs*. They tried to develop ways to prevent or control diseases caused by germs. Medicines were discovered that killed the germs. Remember that a scientist looking through a microscope may have helped to discover the medicine you take when you are sick.

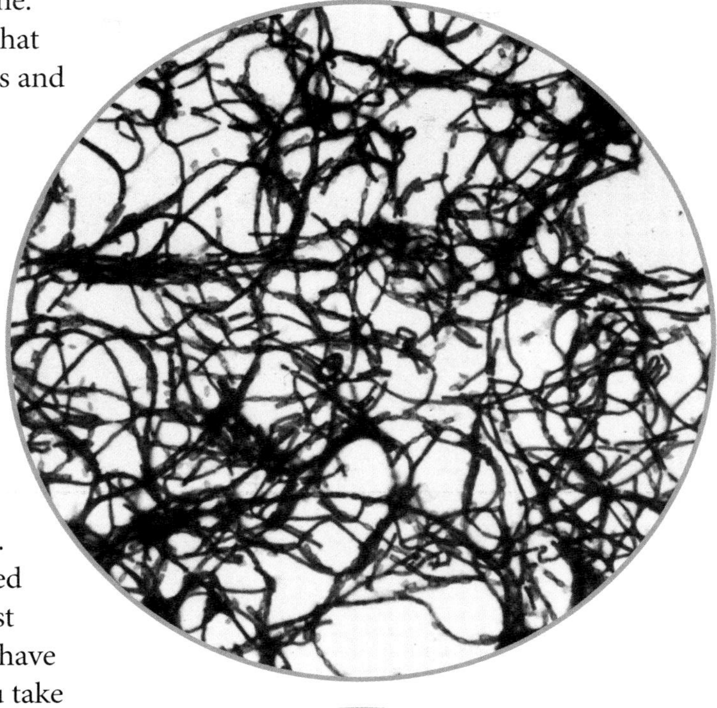

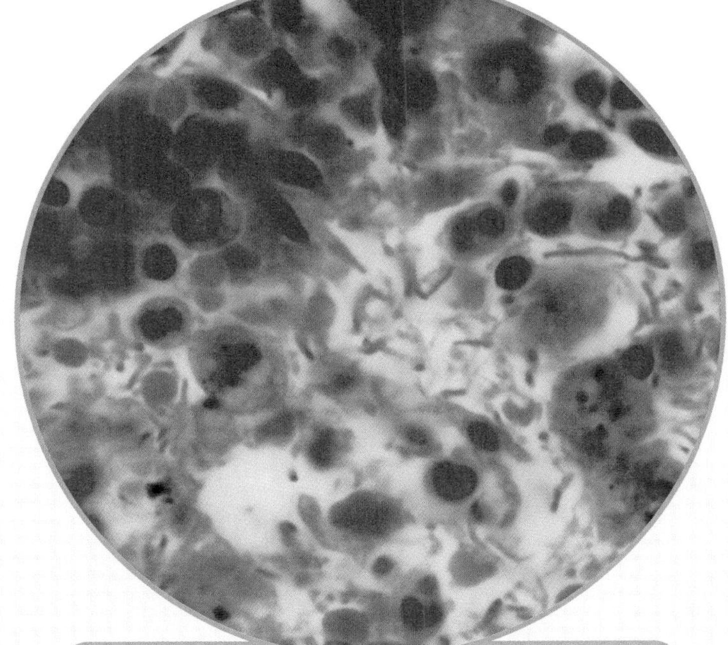

Disease Agent. Scientists discovered that these microorganisms cause the disease known as tuberculosis.

Telescopes and Microscopes

Let's see what you know about telescopes and microscopes.

1. What happens when light hits an object?
2. What is the difference between a convex and a concave lens?
3. How are a refracting telescope and a reflecting telescope different?
4. How are some microscopes like a refracting telescope?
5. Why was the discovery of microorganisms important?

PATHFINDER EDITION

By John Micklos, Jr.

CONTENTS

Cheers rise as the **ponies** splash into the water. It's a July morning on Assateague, an island located off the coasts of Virginia and Maryland. Every year at this time, local cowboys round up the wild ponies that live on the southern end of the island. At low tide, they herd the ponies across a narrow waterway called a **channel** to another island named Chincoteague. Thousands of people come to watch.

The ponies swim across the channel in about five minutes. The crowd cheers again as the animals reach the shore. Back on dry land, the ponies shake the water from their manes. Then they start to graze calmly. Some wander right up to the fence that separates them from the onlookers.

In the Swim. Wild ponies cross Assateague Channel to Chincoteague Island during the annual pony swim. Once ashore, cowboys on horseback lead the ponies to the carnival grounds. Some of the foals will be auctioned there. The auction helps control the pony population.